Standard Operating Procedures for Small Business:

Leveraging Mobile Devices

To our families for their love, support, patience and understanding. You fuel our ability to pursue the passion of helping people.

Forward

This book is intended to help small businesses better utilize a most precious resource, their employees. A small business also stands to increase profits by applying the principles found in this book. This is achieved by standardizing operations, training staff, and introducing principles of Lean Manufacturing and Six Sigma in a simple manner.

The costs in writing a standard operating procedure can be recovered within a couple of months of operation. Using mobile devices such as smartphones and tablets in concert with a simple SOP structure, allows for the writing of high quality documents at a low cost.

Some of the industries that could benefit from a simple SOP program are:

- Food Outlets
- Tradesmen
- Small Manufacturing Shops
- Repair & Overhaul Operations
- Healthcare Clinics
- Self Employed
- Cleaning Services
- Office Transactions
- Start-up Ventures

Introducing Lean Manufacturing principles is done as the SOP is written. These principles include:

- 5S
- Visual Controls
- A3 thinking
- Training Within Industry
- Mistake Proofing

The contents of a short video are more important than the quality of its production. We do not want to win an Academy Award, because we know that the video produced today, could be changed tomorrow by better or more relevant content.

Contents

What is an SOP 8

Numbering System 9

A Word of Caution 9

Link to On-Boarding New Employee 12

The SOP Structure 13

Connection To 5S...14

The Visual Work Place 15

Training Within Industry (TWI)..16

Advantages of Electronic SOPs..17

Picture Process Tables Using Video and Images 18

Synergy..19

Table of Contents 20

Video Shooting Board..21

Label the Plant...22

Leveraging Mobile Devices..23

How to Use the Template..25

SOP Relationship to Improvement Process 30

Section I: Operations 31

1.0 Safety Sections 31

    1.1 Known Risks and Hazards 34

    1.2 Establish the Link Between Safety and Production 36

    1.3 PPE...37

2.0 Record Keeping 38

    2.1 Improvement Opportunity 39

    2.2 Training Aids 40

        2.2.1 Using Excel Review Function. 40

- 2.2.2 To Insert a Comment 41
- 2.2.3 Inserting an Image Into a Comment 42

3.0 Start Up Section 43
- 3.1 Safety 43
- 3.2 5S Activities 43
- 3.3 Supplies and Services 44
- 3.4 Pre-Start Checks 44
- 3.5 Equipment Settings and Adjustment 44
- 3.6 How to Start Machine 44
- 3.7 Complete a Trial Run 45
- 3.8 Critical Question 45

4.0 Process Operation 46
- 4.1 Quality Checks 46
- 4.2 Process Monitoring 46
- 4.3 Disposal of Waste 46
- 4.4 Pausing the Process 47
- 4.5 Ask the Question 47

5.0 Shut Down 48
- 5.1 Stopping the Machine 48
- 5.2 Putting Things Away 48
- 5.3 Prolonged Shut Down 48
- 5.4 Ask the Question 48

Section II: Inspections 49

6.0 Cleaning 51
- 6.1 Hazards When Cleaning 51
- 6.2 How to Clean 51
- 6.3 What to Clean 51
- 6.4 When to Clean 51

6.5 Cleaning Schedules 51
6.6 Ask the Question 52
7.0 Inspection While Cleaning 53
8.0 Preventative Maintenance 55
9.0 Predictive Maintenance 56
10.0 Theory 58
    10.1 Manufacturers Manuals 58
    10.2 Hand Drawn Process Map 58
    10.3 Effects of Material Defects 58
    10.4 Specification Sheets 59
11.0 Error Messages 60
    11.1 Eliminating Error Messages 60
    11.2 Using Talley Charts to Measure Small Stoppages 61
12.0 Troubleshooting 62
    12.1 End State Analysis 63
        12.1.2 Example of an End State Analysis Table 63
    12.2 Manufacturers Manual 64
    12.3 Influences of Specification Variation 64
    12.4 Six Sigma Problem Solving 65
        12.4.1 Belt Levels 65
13.0 Training & Assessment 66
    13.1 Training Screens 66
    13.2 Making Training Structure 67
        13.2.1 Example of a Training Sheet 68
    13.3 Assessment 69
        13.3.1 What is Important 70
        13.3.2 Example of Assessment Sheet 71
    13.4 Linking to Departmental Standards 72

14.0 Revision Log. 73

15.0 Appendix 73

What is an SOP

An SOP is a way of sharing information on how to perform a process or operate a piece of machinery. These documented instructions should be linked to industry regulations, but easily interpreted by any company worker. In most companies, workers are accustomed to having paper copies or a static electronic document that serves as a SOP.

Next Generation Thinking says:

- An SOP does not have to be in print
- Audio Visual (AV) medium is acceptable
- The use of AV opens up the use of other distribution means

An outlined SOP structure allows for AV and other forms of documentation to be used. Like the more popular formats, an electronic format will have to consider the following:

- The AV will need to be stored and controlled
- The SOP can must refer to AV
- The SOP must contain relevant information to safely operate the process

Similarities between paper and electronic SOPs:

- Must be easy to write
- Controlled
- Easy to understand
- Contains input from all parties

Numbering System

We recommend that a decimal number in system be used

First digit section number		1.
Second digit	Subsection	1.1
Third Digit	Subsection	1.1.1
Letter	Table Ref	1.1.1.A

This allows for any section of the SOP to be easily referenced. A big advantage is the use of references to shoot short video clips

Example:

    0.2.B   Refers to section in table marked "bag filler" two pages on

A Word of Caution

Writing an SOP may be viewed as a means of improving a process. However, if the process is flawed, merely documenting the process will not improve it. Managers may view the problem as staff not following procedures and believe that documenting the process will improve the result. Staff operate the plant in different ways (See Figure A), because there is variation in the system and also in the way we communicate expectations.

Examples

1) Bottle Labeller Issues
i. Labels not applied to bottles straight
ii. Staff operated machine differently
iii. Buying a new machine did not solve problem
iv. True cause was that bottles were not straight

There was no way of correctly operating the machine to see a straight label

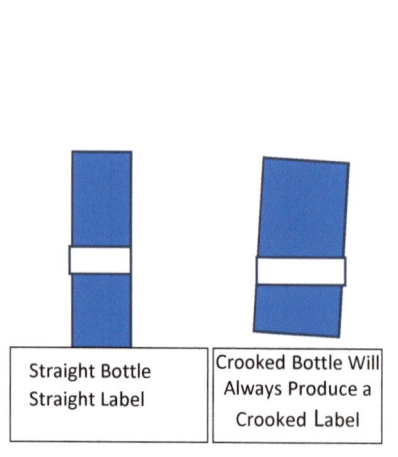

Straight Bottle Straight Label

Crooked Bottle Will Always Produce a Crooked Label

2) Bag Filler
i. Bags did not seal properly
ii. Staff operated plant differently
iii. Happened on different brands of machine
iv. True cause was plastic liner incorrectly glued, causing bags to deform randomly when filled.

Results: The deformed bags would not seal consistently

3) Metal in Cheese

i. Small nuts and washers became loose and entered product. They could not be detected by the metal detector, which was placed after a Chiller holding 5,000 blocks of cheese.
ii. Elaborate procedures and training were put in place to mitigate and perform trace backs
iii. **Solution:** Nuts and bolts made bigger, so they were easily detected by metal detector, which was place before the chiller.
iv. **Results:** Any loose nuts and bolts were immediately detected. Procedures were simplified.

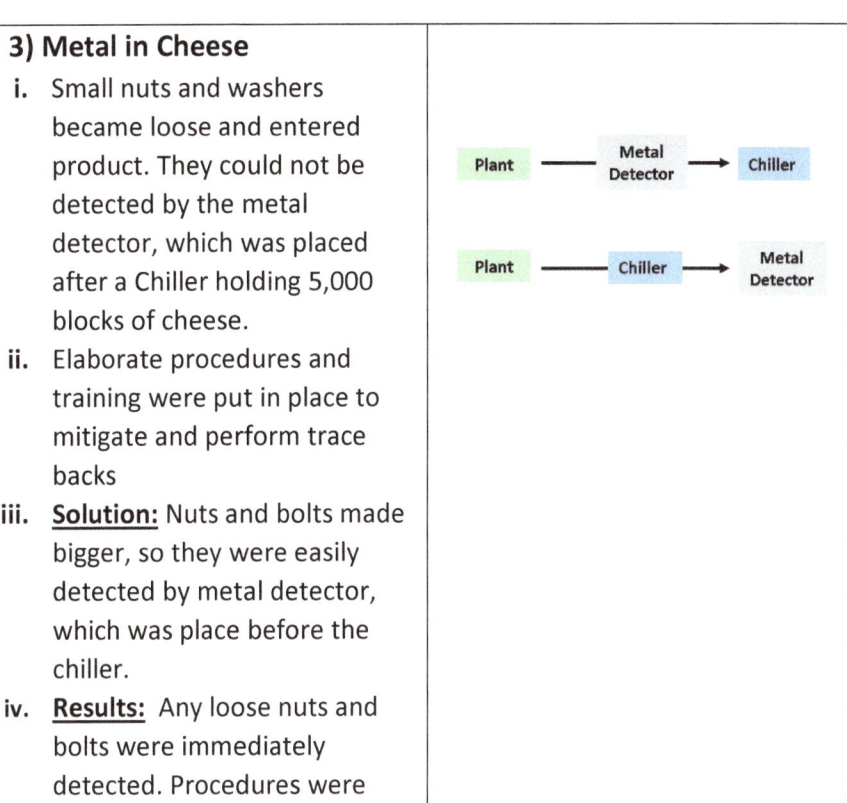

Figure A

Link to On-Boarding New Employee

On-boarding is the first stage of training a new employee and is a critical first impression. However, the brain will only remember six or seven things during this process.

A Good SOP Will Fill in the Gaps

- The safety section of the SOP should reintroduce what is presented at the on-boarding event to reinforce learning
- Safety links specifically to the machine or procedure should be identified and studied.
- It will focus on the essential items or the "must know"

The Musts

- Trainees must be given something to refer to after they have forgotten some of what was instructed. In some cases, providing a manual with the new hire's name printed on it will generate ownership.
- They should be told where and how to find information
- Train staff in using reference material
- Periodically perform spot surveys to test what an employee knows and if they can find information that they have forgotten

The SOP Structure

The following SOP structure is recommended:
- 1.0 Safety
- 2.0 Record Keeping
- 3.0 Start Up
- 4.0 On Product
- 5.0 Shut Down/Stopping
- 6.0 Cleaning
- 7.0 Inspection While Cleaning
- 8.0 Preventative Maintenance
- 9.0 Predictive Maintenance
- 10.0 Theory
- 11.0 Trouble Shooting
- 12.0 Error Messages
- 13.0 Training & Simulated Screens
- 14.0 Revision Log
- 15.0 Appendix

The number of items in each section will vary. Regardless of the task, product or industry, the SOP will contain reference to the above. It will not matter if it is a laboratory process, office transaction or a machine maintenance, the same sections will be needed. The importance and size of each section will vary between industries and applications.

As the process improves, the SOP can be updated, and others can share the improvements. Using video content, makes writing an SOP easier. Low cost video of a procedure meets most requirements. It is essential that the video be documented and controlled.

Connection To 5S

5S is focused on organizing the workplace, so it is an easier, safer and more productive place to work. People need to be shown how the workplace should be organized, relative to the work they do. Workers know what needs to be done, but do not have the means to take action. Leadership has the means to address issues, but do not know what to address.

5S Competencies
Sort
- Remove all items that should not be there
- Put a system in place to prevent items from reappearing
- Make regular management spot checks
- Make worker improvements
- Eliminate sources of dirt and contamination

Set in Place
- Get those who work in the area to decide where things should be kept
- Label where things should go

Shine
- Give the place a good clean
- Make it easier to clean
- Make cleaning improvements

Standardize
- Introduce worker driven cleaning schedules
- Define what should be inspected while cleaning
- Make predictive/preventative maintenance routes for workers

Sustain
- Make 5S a KPI of the business
- Get the team to make visits to other factories to gain ideas and promote those who support 5S (benchmarking)
- Introduce a reward and recognition system

The Visual Work Place

 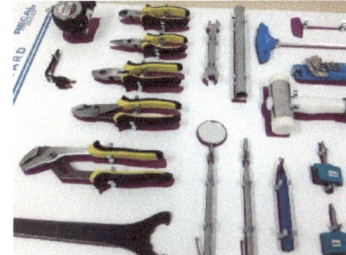

http://www.georgekk.co.uk/shadow-boards/tool-boards/

Figure B – Shadow Boards

1) There is no need to train customers on how to use the supermarket
2) There are no significant safety concerns
3) Everything is in a place that makes it easier to find

https://pixabay.com/en/photos/supermarket/

Figure C – Supermarket Example

Though the gas station has potential hazards, these have been removed by organizing the workplace. There is no reliance on procedures.

This workplace is safe by design

https://pixabay.com/en/gas-pump-petrol-stations-petrol-gas-1914310/

Figure D – Fuel Station Example

Training Within Industry (TWI)

During WWII, a training system was put in place in the United States. It took poorly skilled, less able workers and trained them to become a highly skilled and productive workforce. The system focussed on making the trainee feel at ease, broke the tasks into learnable sections and trained by doing instead of reading. The mantra of a system such as this is, "If the learner has not learned, the trainer has not trained correctly".

At the end of every session, the following questions are asked, "Is there a better way of doing this?" "Is there a safer way of doing this?"

With TWI, there was a prescribed way of doing things and everyone was trained in that method. They were expected to follow the method or procedure until a better way of doing things was found, and staff were retrained in the new procedures. The best way was never set in stone. At the end of WWII Japan adopted a similar system.

The questions that must be asked
- Can this be done safer
- Can this be done better

The improvements get added to the SOP and shared with others.

Keeping the SOP up to date is essential.
Using an electronic form has advantages.
- A tablet or other device can be kept at the workstation.
- This can be updated wirelessly.
- It encourages the concept of using reference documents.

Advantages of Electronic SOPs

Smartphones, tablets and other devices have introduced a more efficient way of doing things.

- SOPs can be easily written
- SOP can be easily updated
- We can provide staff a device to learn from
- The device can contain video and many other media types
- The SOP can easily be translated into the native language of the learner.
- Video technology means that almost anyone can contribute to an SOP in an organization

https://www.publicdomainpictures.net/en/view-image.php?image=168789&picture=cave-man-chiseling

Takeaway: Low cost publishing is very achievable

Picture Process Tables Using Video and Images

These are a highly effective method of delivering information. It is a combination of image and the written word. Think of comics

Example

1) Make a Table i. Insert table	
2) Select a Style Heading i. This will always give the same font	
3) The Style Font Can Be Modified i. Right click on it ii. Click on Modify iii. This will bring up other menus	
4) Add a Border and Numbers i. Insert a Box ii. Make a number list	

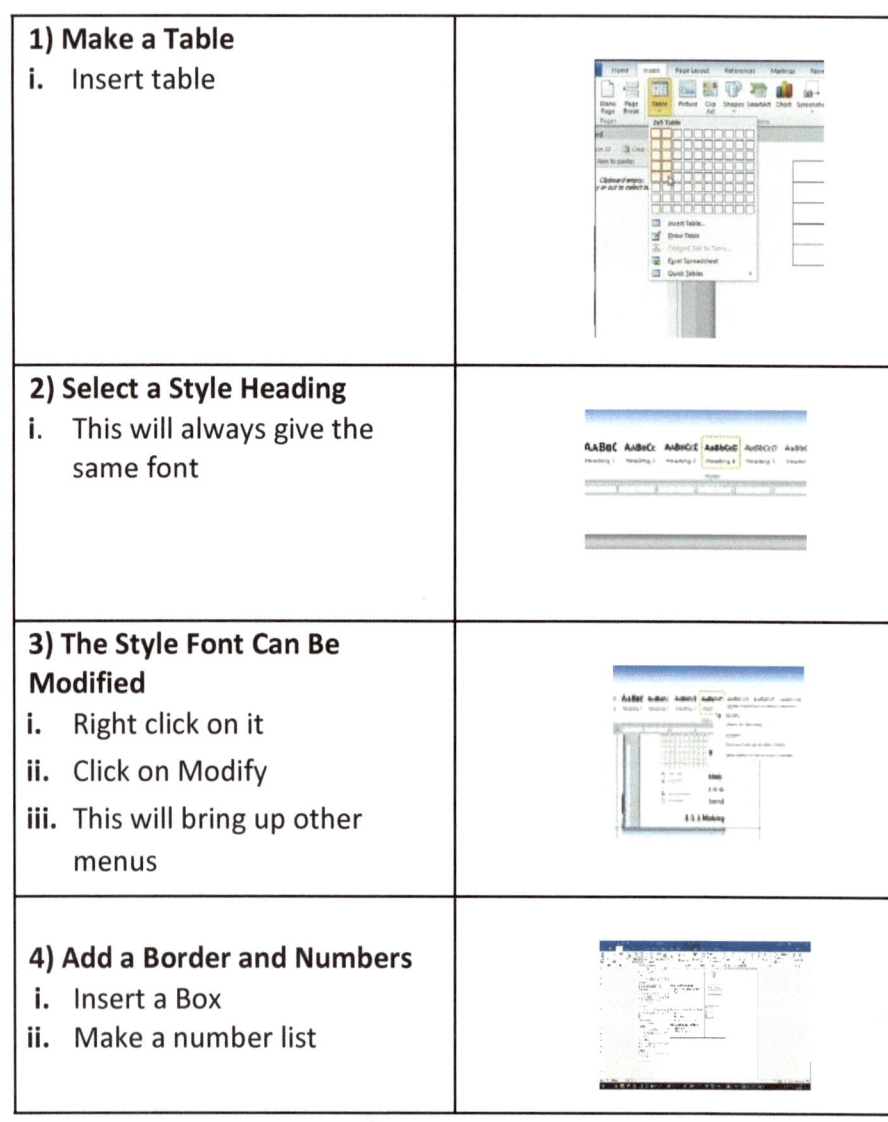

Figure E

Synergy

Why Use Synergy?
- Each computer can control its own program independently of the others reducing the load on each computer, so they run easier
- Items can be cut and pasted and transferred between computers with the same ease as a second screen
- Capture devices can be used independently
- Any device can be fed into one of the computers. The data can then be very easily transferred to the second or third computer

What Can be Accomplished?
- We can take a video of the process
- Feed it into one computer
- Listen to the commentary
- Take screen dumps and populate the picture side of the table
- The process is recorded in the order it is performed
- Photos directly linked to the writing
- Headings which can be used as a table of contents
- The table of contents can then be made into a training and assessment plan.

Alternatively, contact a company like https://opexperts.org/ and they will arrange to make an SOP from the video you record.

Table of Contents

A table of contents is essential

- It allows users to easily find material.
- It can be easily updated if the content changes
- It can be copied and used as a training plan
- The level feature of MS Word allows easy control of how much detail is in the table of contents
- Pages of the SOP can be located via a click of mouse

Note: There are many videos that provide DIY instructions on creating a table of contents

Tip

Activating hyperlinks allows for navigation in mobile devices

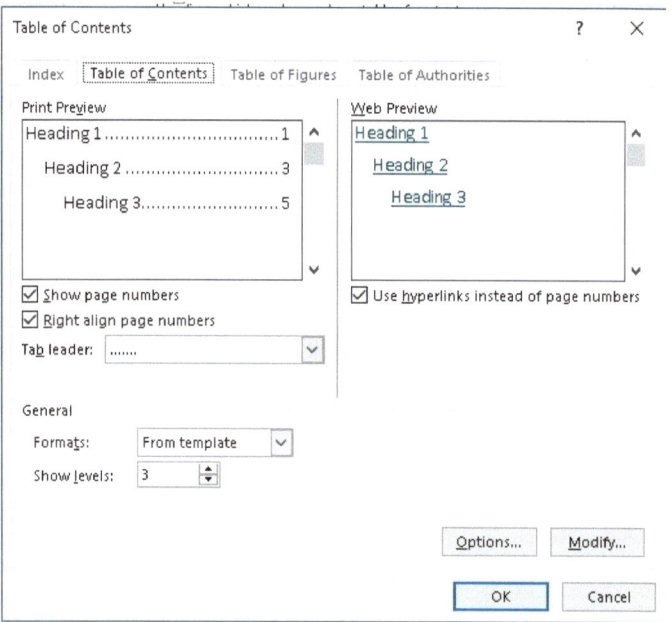

Figure F

Video Shooting Board

1) Make a Board for Video i. Title of SOP ii. Section of SOP the video refers to iii. Date shot & version number	"SOP Writing." 2-13: Making Video SOP Date: 13 Oct 2014 V1c
2) Shoot Video Starting with Title Board i. Hand written titles are OK ii. We are not shooting a blockbuster with $30m budget iii. It's just to share information	
3) Edit and Save as a Controlled Document i. Video Screen capture is great. ii. Allowing captions can be used if sound quality is poor iii. Use specialized microphones if good sound quality is required iv. Low cost microphones plug into smartphone	

Figure G

Label the Plant.

It is absolutely essential to label the plant. An unlabelled plant is like a city with no name or roads with no street numbers

- For productions screens, label the plant to match what is on the production screen.
- For pipes use international labelling conventions. These can be found by completing a google search.
- Giving parts that are similar nomenclature is useful.
- If we have a plant named 'Plant 1', 'Plant 2' and 'Plant 3', it may be easier to refer to them as:
    - Kermit (Plant 1)
    - Miss Piggy (Plant 2)
    - Cookie Monster (Plant 3)

The Blackboard Menu

https://commons.wikimedia.org/wiki/File:Concrete,_WA_-_5bs_Bakery_-_partial_menu_01.jpg

Figure H – Handwritten Example

Leveraging Mobile Devices

The Five Basic Configurations Using a Template:

1) Print the Template and Hand Write Content

This is very useful for businesses with under 10 people. In effect, hand-written notes become an organized "Knowledge Book." Procedures can be quickly written, updated and standardized, Issues associated with using a computer are avoided. Document control may not be important, as there may only be one copy of the SOP that it is kept where it is needed. The hand-written notes can be given to a data entry person if required. Hand written SOPs are equally valid as computer written ones as they both contain the same information. The hand-written notes can be made directly at the workplace.

2) Load Template Directly to Computer

This allows users to shoot a video. The video is loaded into the computer and then edited and loaded into SOP. The SOP is written on the computer. The disadvantage to this method is that the writing portion of the SOP is accomplished at a remote location.

3) Load Template into Laptop Connected to USB Camera

This allows video to be loaded into the laptop and edited. The advantage is that the SOP can be written at the workplace as the process is taking place. The video can be reviewed and edited at the point of process.

4) Load Template Directly into Mobile Device

Video and text are edited inside a mobile device. The advantage is that most mobile devices have video, text, and audio processing available via an app. Owners of mobile devices are often highly skilled in their use.

5) Connect Computer and Mobile Device with Your Favorite Program

This allows the SOP to be edited by both computer and the mobile device. This combination is a powerful tool and allows the best of both systems to be used. There are many file sharing applications that allow a computer and a mobile device to communicate to each other.

Connecting devices to a computer has many advantages.
- Editing can be done on both computer and mobile device
- The SOP is available on a mobile device or multiple devices

How to Use the Template

1) Print the Template and Hand Write Content

A printed copy of the template is highly recommended for all methods for use. All video production relies on the use of a story board and a clip board to help organize the process.

In many cases, the information that is needed for an SOP is already known. The problem is that it is not organized. The SOP structure allows for the organization of material. It shows that there are other activities outside of the process that need to be considered.

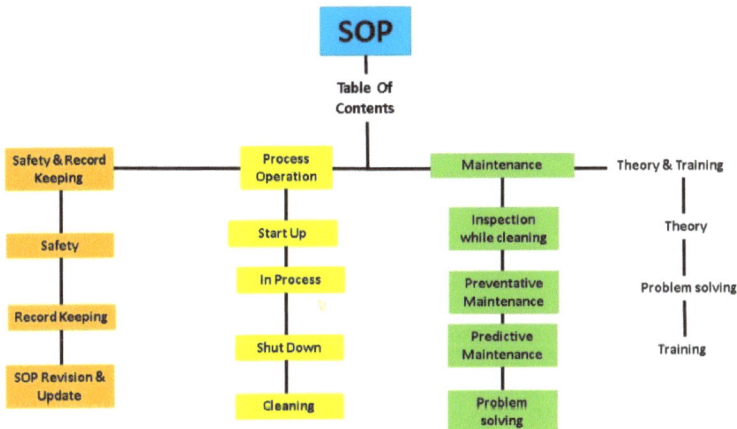

Figure I – SOP Structure

The Hand Written SOP

- Can act as a check list.
- Highlights areas for improvement
- Captures improvement

These can be very effective for casual workers. They outline items of importance, such as:
- Emergency procedures
- Safety
- Cleaning

> **Tip**
> Save hand written SOP in drop-in folder. Pull a page out and replace if changed.

2) Load Template Directly to Computer

Print the SOP template. This becomes a storyboard for the video. All video and movie production uses a storyboard. Every page of template has pop up comments that can be hidden. The reference number links to the appropriate section of the book.

In this example 1.1.3 and 1.1.4 has explanations of how to populate the SOP. Instruction on how to add comments are explained in section 2.2.2 of this book.

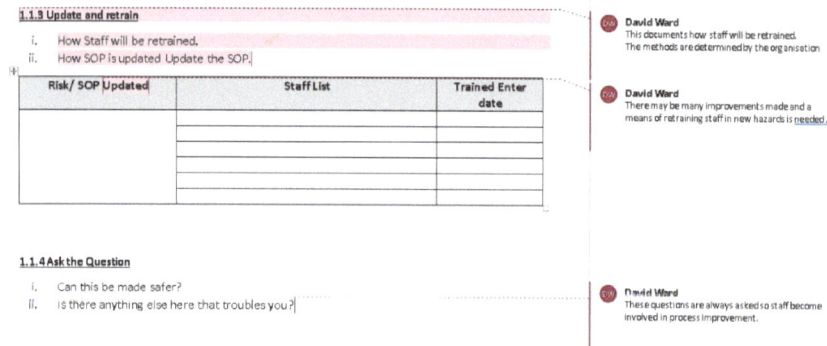

Figure J – SOP Template

Links are added to video

Activity	How to Video
1) How to write SOP i. Shoot the video ii. Edit the video and provide link iii. Edit supporting text	<u>(Click to open address of video)</u> <u>Click to follow link open Video</u>

Link to existing documents

If there is an existing document or procedure provide a link to it or paste it in the appropriate place

Tip
Name and save the video with a nomenclature that matches the SOP (i.e. 2.1.1 A). Write the nomenclature in the printed-out template. This makes it easier to find the video.

3) Load Template on Laptop Connected to USB Camera

Using a USB camera allows video to be directly recorded and saved into a location. This simplifies the process and video can be edited at the point of shooting

Activity	How to Video
1) How to Record Video i. Record the video ii. Edit with laptop editor iii. Edit supporting text iv. Name and save	

Great Time Savers

Connecting a second computer via Synergy is a great time saver.

- This allows one computer to be used to edit and the other to be used to write.
- The same mouse controls both computers
- It eliminates communication by saving and, then re-opening from a file.

Choose a Video Camera with On-board Editing

- The video can be reviewed after shooting
- This allows the video to be edited as it is shot
- The final product can be reviewed by those "on the job"

4) Load Template Directly to Mobile Device

There are a very large number of mobile devices. The specifications of the mobile device will determine how easy it is to use. There is however, external hardware and apps that will improve the performance of the mobile device

Microphones

These include lapel microphone, contact, directional, omni, and shotgun. The use of such microphones dramatically increases audio quality and decreases the need for audio editing.

Audio Editors and Filters

These can be used to filter the sound being recorded.

Video Enhancers

These are added to increase the quality of the images being shot. They vary, dependent upon the make and platform of the mobile device.

Infra-red Attachments

These can convert the smartphone to a very effective maintenance tool.

External Storage Devices

These increase the storage capacity of the smartphone. Some can be configured so video is recorded directly to the drive. Some have the option of being transferred to a computer.

Text Editing and Text Entry

There are numerous apps that make it easier to enter text into a smartphone. The make and platform of the mobile device will determine which is best to use.

Selfie Sticks

The can be used for taking photos remotely. They are also useful in performing maintenance checks.

5) Connect Computer and Mobile Device with your Favorite Program

There are numerous ways to share files between devices. Below are a few examples.

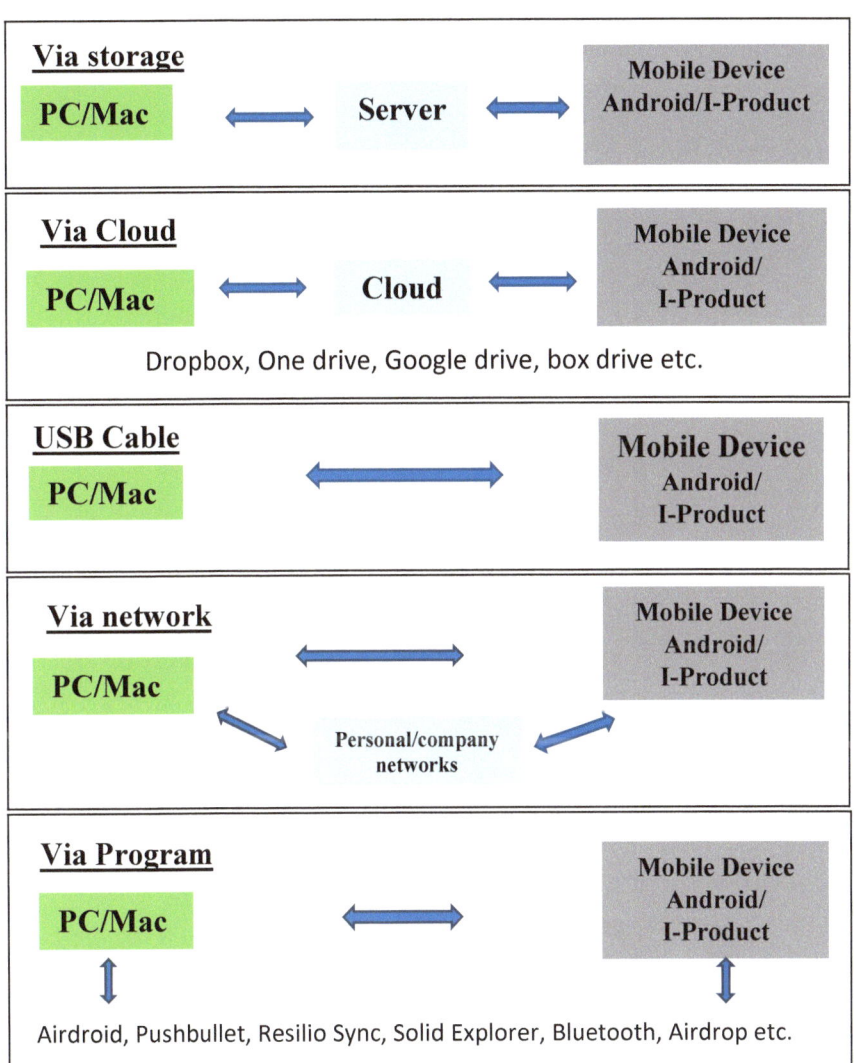

Figure J – File Sharing Techniques

SOP Relationship to Improvement Process

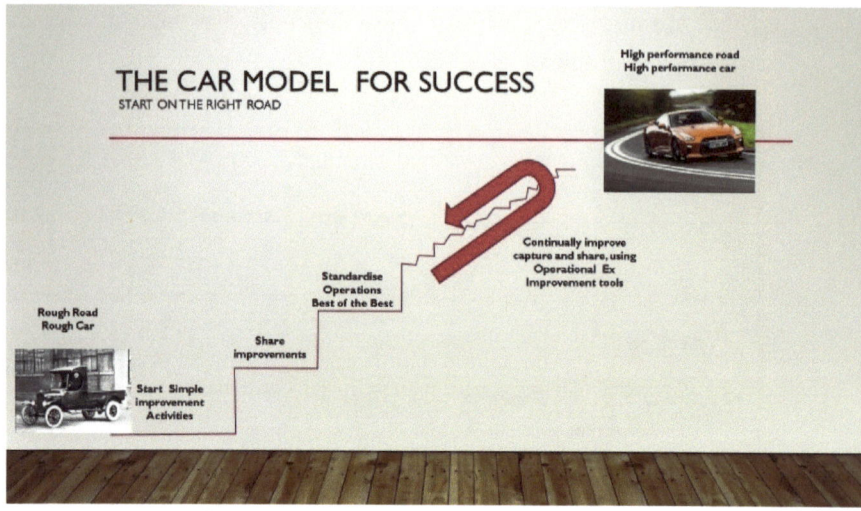

Figure K –Car Model for Success

Starting Too Soon

The process is unstable and full of variation. If advanced statistical analysis is introduced at this stage, it will fail because there is too much variation in the system. It's like driving a highly tuned car down a rough road. The road will destroy the car. A start needs to be made on making small practical improvements. These will deliver large improvements over time. An old Model T is most suited. It will not be fast, but it will be reliable.

Starting Too Late

The process will never be standardized, and improvements will fail. The Model T is not suitable for using many lean manufacturing tools. It will be too slow and use too much fuel. The sports car is more efficient.

The Optimum Starting Time

A good time to start is shortly after making improvements. Staff involved in making improvements will want the improvements sustained and will have "Buy In."

Section I: Operations

1.0 Safety Sections

There are many good reasons why we should make safety the number one priority. Below is a comparison of the differences that we see between how low performing companies and world class companies see safety.

Low Performing Companies

- The worker intervenes to keep the plant going and maintain production
- The same small stoppages continue to occur
- This can continue for a period of time, and often there is an incident where a cause is finally eliminated
- Production has been continually lost until the cause was fixed
- The incident causes losses

World Class Companies

- Fix the item the first time it happens
- They do not lose production with small stoppages
- There is no loss associated with an incident
- Have engaged workers so stoppages are predicted and prevented

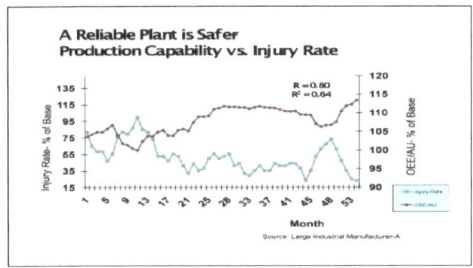

Figure 1.1 – World Class Performance

Embedding Quality

"We always find time to do it again, but never seem to find time to do it right."

The statement above comes from a myopic focus on production. Staff will try to get maximum output at the expense of quality. However, poor quality is worse than under producing.

Not only can a product that contains defects not be used, it also needs to be disposed of. The same product must then be made again.

A solution to avoid the previously stated risks is to focus the content of the SOP towards doing it right and doing it safely, the first time. Process inputs that affect quality should be included in the SOP. The troubleshooting section should show how to rectify faults. In the event that process inputs are not understood, contact an operational excellence group such as http://Opexperts.org/ to begin identifying and including this valuable information.

Focus on Safety – Rate, Safety and Quality always increase
Focus on Quality – Rate, Safety and Quality always increase
Focus on Rate – Rate, Safety and Quality always DECREASE!

Two Key Concepts of Hazards

1) Be Positive
Always state what *should* be done.

Do not touch machine is remembered as touch machine because the brain usually remembers the last section of an instruction.

Staff will be left wondering and think "what should I do?"

A better instruction is "when inspecting machine, stand behind the line." Stand behind the line is what is remembered.

2) Limit Information
Too much information is counterproductive. The following is a safety instruction for a vacuum cleaner:

Do not leave unattended, Do not use outdoors, Do not allow it to be used as a toy, Do not allow children to use machine, Do not use with a damaged cord, Do not pull by cord, Do not pull around corner, Do not run appliance over cord, Keep cord away from heated surfaces, Do not unplug by pulling cord, Do not handle appliance with wet hands, Do not put objects into openings, Do no used with openings blocked, Keep hair away from moving parts, Do not pick up anything that is burning, Do not use without a dust bag, Turn off before unplugging, Be Careful on stairs, Do not pick up combustible liquids, Do not use in confined spaces, Unplug from wall before cleaning, Keep all chemicals out of reach of children, Do not use pesticides.

There is too much information to be absorbed as a safety lesson in this example. Our approach is to introduce key points. Everything else is introduced and trained at the time it applies.

1.1 Known Risks and Hazards

1.1.1 Document Risks, Incidents & Near Misses

The risks identified should be re-introduced later in the SOP. This is a check list to ensure that all risks are presented and appear at the same time as the relevant task is taught. A simple table is what is needed. The table below is a good example of how complex instruction can be reduced to an easy to understand directive.

Risk	Represented
Traffic Hazard F1(Fatal Risk 1)Video or Document Link	3.1.1 Disposal of Trash 6.1.4 Cleaning
Incident Video or Document Link	4.1.1 Position of E Stop 5.1.4 Shut Down

Table 1.1 – Simplistic Instructions

As an example, the vacuum cleaner risks would all be placed in the SOP and taught as they arise in the process. The key item we want staff to remember is, that before someone works on a machine, that it should be unplugged it from wall. This is constantly reintroduced, because it is a fatal risk.

1.1.2 Ask the Following Questions
- Can this be made safer?
- Is there anything else here that troubles you?
- This immediately places the responsibility on the staff member to take responsibility for his or her own safety

1.1.3 Update and Retrain

The company should update SOPs as the need arises. Retraining of the updated SOPs should immediately follow a new release. During each training event, the applicable staff should always be asked how to improve the process.

Items to Include in Updated Retrain Log:

- Record staff names and date of training
- Update the SOP.

Risk	Staff List	Trained Enter date
Traffic Hazard F1(Fatal Risk 1) [Video or Document Link](#)	Jo	
	Mary	
	Peter	
	Jane	
	Belle	
	Alberto	

Table 1.2 – Training Log

1.2 Establish the Link Between Safety and Production

As a company, we prefer to have no stoppages. Avoiding risks always improves production and reduces losses

- Find a place where production may stop
- Ask the trainee "if we take a shortcut to get the plant running what will happen tomorrow?"
- Ask the trainee "which is better, having no stoppages or taking shortcuts?"

Example

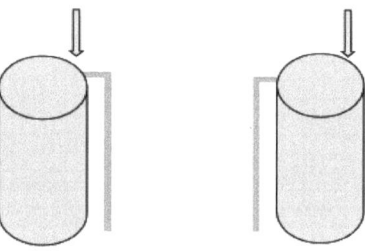

Figure 1.2

Situation
The inflow pipe was next to the overflow pipe. Whey concentrate spilt to the floor causing a slip hazard
Improvement
 Move the overflow to the other side of tank
Cost to Implement
$150
Saving
400,000 liters of water
$150,000 in Whey losses
250 hrs per year in cleaning time
Takeaway
Eliminating risks usually saves money

1.3 PPE

A valuable advantage of using electronic medium is that it is easy to record video showing correct wearing of PPE. Images or video can be used to demonstrate:

1) What PPE Needs to be Worn?
i. Ear Protection
ii. Safety Boots
iii. Hair Nets
iv. Respirators
v. Face Shields
vi. Gloves
vii. Etc
2) What Harm Does PPE Prevent?
i. Ear Protection - Against hearing loss
ii. Safety Boots - Against crushing
iii. Hair Nets - Against entanglement and are also for product safety
iv. Gloves – Bloodborne pathogens
3) Does PPE Provide Complete Protection?
i. No, because if it can become ineffective, become damaged, not fit correctly, or not worn properly
ii. It is best to eliminate the hazard so PPE is not needed
4) Demonstrate Proficiency
i. Add an assessment question that trainees are seen to be wearing PPE

<u>Table 1.3 – PPE Considerations</u>

An example of this is seen when hospital disease and infection control departments conduct N95 fit mask testing in accordance with OSHA standards. There are requirements that advise against chewing gum and even a variation in how the mask is placed on a bearded face. The procedure takes 10-15 minutes and must be done correctly.

2.0 Record Keeping

Often an assumption is made that staff know how to fill out forms, though they have not been trained in how to do so. Factors that affect the ability to fill in forms include:

- Literacy issues
- Language differences
- The forms are poorly designed
- Information is recorded that is not required

The followings steps are useful:

- Review the information on the form
- Ask what the information is used for and what would happen if it were not collected
- Investigate redesigning the records and if some fields can be pre-printed. Examples would be date, shift number, etc.

A good training example:

- Have managers fill out one of the forms and time how long it takes to complete
- Calculate how many times a year the form is filled out by how many people
- Multiply these figures
  3 min x 10 people x 300 times = 150 Hrs per year
- Find the total number of forms to be filled
- The total time spent filling in forms identifies

2.1 Improvement Opportunity

Improving record keeping is often a missed opportunity for improvement. Most operational excellence professionals are trained to walk the current state of a process to understand where the failures might be located. It becomes extremely critical to have the latest information available about how a process is to be performed, so the waste and sources of variation stand out. On another note, regulation agencies in a lot of businesses require that documentation be correct and up to date. Loss of compliance certification or even financial impacts can be levied against companies that do not emphasize proper record keeping.

Record Keeping Tips
- Make a list of all records that need to be kept
- Provide a copy of a correctly fill out record
- Create a check list for training staff

A Practical Example of Improvement

A cheese plant had employed a worker in a low-level position for 5 years at a basic wage. When the new training system was introduced, the question was asked about forms:
"Can this be done better?"

His answer was "Do everything in MS Excel". The employee evidently was an MS Excel expert and was studying computer programming. He set up the site forms, so that data was entered in one point and automatically sent the information to 15 other forms. Users received pre-printed forms with all the correct information prepopulated.

The result of this effort was that the forms were updated 15 times in one year. One department reduced the number of forms from 7 down to 2. This turned out to be the start of a culture of improvement for the plant.

2.2 Training Aids

Microsoft Excel can be used to create a form. Comments can be used to provide an explanation of each section

2.2.1 Using Excel Review Function.

Hovering over a cell displays a comment

Images can be inserted

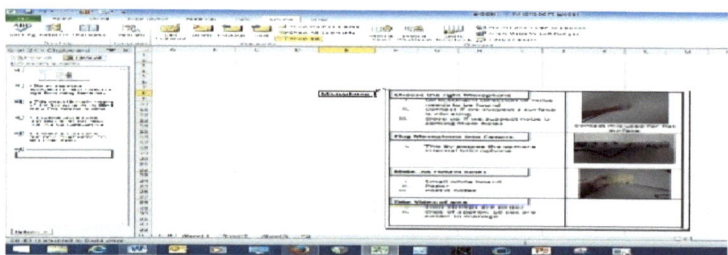

Figure 2.1 – Image Insertion

- Though simple this is a very powerful tool
- Allows interaction
- Slashes description functions
- Links to video can be made

2.2.2 To Insert a Comment

1) Click Review i. New Comment	
2) A Box Will Appear i. Type message in the comment box	
3) A red Mark Indicates a Comment i. Hovering will display the comment	
4) Right Click in Cell i. This brings up the edit box ii. Selecting edit comment allows t comment to be changed	

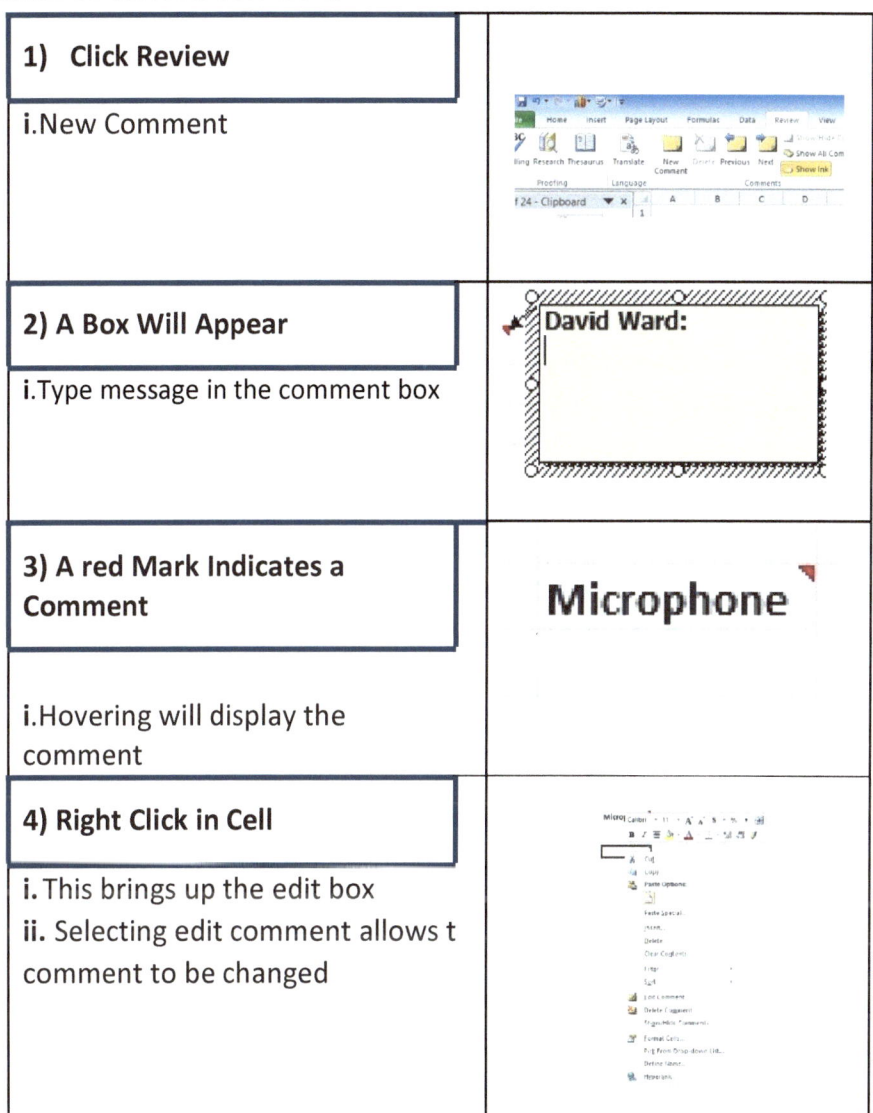

Table 2.1 – Comment Insertion Instruction

2.2.3 Inserting an Image Into a Comment

1) Right Click the Cell With Comment i. Select "Edit Comment"	

2) Right Click Border of Comment
   i Border may need to be clicked twice.
   ii Select Format Comment	

3) Select Color and Lines i This will display another box	
4) Select Fill Effects i Select the drop-down box colour	

5) Select Picture
   i The insert box will appear
   ii Insert Press **"Ok"** till finished	

6) Scans or Other Items Can Be Inserted i Very helpful if the form requires knowledge of procedures	

Table 2.2 – Image Insertion into a Comment Instruction

3.0 Start Up Section

The start-up section includes a callout of the necessary steps to begin a process. Often times beginning a process will include tasks such as warming up ovens, counting sterile instruments for surgery or measuring initial chemistry in a processing tank.

3.1 Safety

Identify the Hazards Associated with Starting the Machine

Ensure that any manufacturer warning about equipment start-up are identified and included in an SOP. Such warning could include electric shock, confined space or exposure to chemical in start-up mode.

Demonstrate How to Stop the Machine in an Emergency

The SOP should include a picture of emergency stop switches or detailed information on the safe shutdown sequence of equipment

Show How Machine Can Be Isolated

It is important for an SOP to show the detail of how a machine can be isolated from heat, electricity and fluid flow through ducting, switches or valves

3.2 5S Activities

Set in Order

Any equipment such as tooling, keys or measurement equipment necessary for start-up should have a dedicated location that is tracked on the facility 5S auditing

Putting Things Away

The SOP should provide details on when it is appropriate to put start-up items back in their home location

Workplace Organization

The work area should be organized in such a manner that all items necessary for start-up are always visible and available, so that the process may start on schedule.

3.3 Supplies and Services
Confirm that Material and Services are Available
The SOP should include quality checks and any relevant specifications. It should also make reference to ensuring that any support organization are up to speed on how and when start-up processes are conducted. For instance, it is critical to first shift (typically high volume) operations that a maintenance organization opens up facility air regulators after the lower volume third shift has ended.

3.4 Pre-Start Checks
A standard operating procedure will clarify what needs to be checked prior to production or the start of a process. For instance, it is critical for a scrub technician to count instruments in front of a surgeon prior to and after a surgery, so foreign objects are not left in a body. This is the same principle in machine assembly, particularly in the aircraft business. Foreign object debris can cause serious damage, most of which can be prevented by including pre-start checks and a good mistake proofing culture.

3.5 Equipment Settings and Adjustment
Whether in an operating room setting or industrial machine shop, equipment adjustment is critical to yielding a predictable result. Instructions in an SOP should provide such details to standardize items as cutting tool placement or temperature settings on a reflow oven.

3.6 How to Start Machine
Equipment operation in an SOP should be designed to help the user navigate through a start-up procedure according to how the supplier directions. The original equipment manufacturer will generally provide a manual that includes operation instructions for the equipment. Many times these manuals are provided in an electronic medium such as a pdf file. Electronic

operation manuals can be linked directly into an electronic SOP easily. Note that paper copies of operating procedure can be scanned into pdf format as well, making that a viable solution. In addition to linking pdf files into an SOP, video can be taken of the operation of the equipment, providing a great source of training and reference material for all levels of staff.

3.7 Complete a Trial Run

It is a best practice to run a trial of the SOP after it has been drafted and before it is finalized. Another great idea is to have process improvement, process engineering, facilities, maintenance and even the supplier onsite (if feasible) to witness the SOP trial. This group of cross functional folks can help make decisions as to how many trials must be run to satisfy the safety, reliability and sustainability of the SOP. They should also discuss potential risks that might not be addressed in the SOP. If it doesn't make sense to add a countermeasure for a certain risk, then this focus group can discuss how best to handle the risk and what corrective action to take.

3.8 Critical Question

As is the case with designing a new product, process or service, we must ask the following question continuously of folks that use the SOP daily as well as for training material:
- Can this be done easier/better?
- Can this be done safer?

4.0 Process Operation

The following headings are the actions performed while running the machine or the process.

4.1 Quality Checks

While a process is being performed, it is helpful to ensure that critical characteristics of the product or service are being met. This involves checking to ensure that what is being produced is in specification (voice of the customer).

4.2 Process Monitoring

Critical characteristics are the items that need to be checked while a process is running. In the manufacturing world, these can be material hardness or the measurement of a physical dimension to the blueprint. In healthcare, this might be a pinnacle event that would cause harm to a patient. The inputs that would drive either bad outcome should be monitored so that early intervention can occur. An SOP should provide direction on when and how to take these key measurements and what the response should be. Whether the measurement method be a periodic audit or statistical process control, details must be understood in the SOP to provide value of this costly inspection.

4.3 Disposal of Waste

The SOP should direct anyone who has the responsibility of waste disposal whether that be bio-burden or machining chips the following information:
- Where does waste go?
- List recycling procedures

4.4 Pausing the Process

Periodically, a process must stop to accommodate planned or unplanned events. The SOP must address the following:
- The procedures for pausing production
- The procedure for restarting the process if it is different than the start-up called out in section 1.0 of the SOP

4.5 Ask the Question
- Can this be done safer?
- Can this be done easier/Better?

5.0 Shut Down
This describes what activities need to be completed when machine stops

5.1 Stopping the Machine
There are many processes and machinery that require a certain sequence of shutdown. For instance, on rotating machinery testing equipment, a torque cannot be suddenly released, or it will cause major bearing damage. In the event that the process shutdown cannot be automated, the SOP should outline how the process should be shut down.

5.2 Putting Things Away
This section of the SOP should give direction on what to do with any item that was used in support of operation. This might include direction on which tool location fixtures should be returned to.

5.3 Prolonged Shut Down
There are many processes and machinery that require attention even when they are not being utilized. An example of this would be gearboxes on test fixtures. If the process will not run for a prolonged time, output shafts may need to be covered or the gearing turned periodically to circulate lubrication and reduce the chance of corrosion.

5.4 Ask the Question
- Can this be done safer?
- Can this be done easier/better?

Section II: Inspections

6.0 Cleaning.
7.0 Inspection while cleaning
8.0 Preventative maintenance
9.0 Predictive maintenance

We recommend a broad approach where Schedules define:

- When it is done
- Who does it
- What is done
- The definition is usually defined in schedules

What	Day 1	Day 2	Day 3	Day 4	Day 5	Day 6	Day 7	Day 8	Day 9
Action 1	John							Shift 1	
Action 2				Mary					
Action 3		Peter							
Action 4						John			
Action 5									
Action 6				Jane					
Action 7							Sue		

Table 5.1 – Inspection Matrix

Mobile devices allow links to be used to show how an action is done. Clicking on an action links to a screen that show how an action is done.

Big Advantage.

If workers are unsure of an action they can refresh themselves by linking to the section that describes how it is done.

| Maps define where | OPL define how and why |

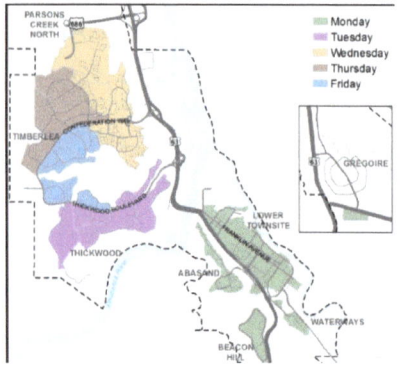 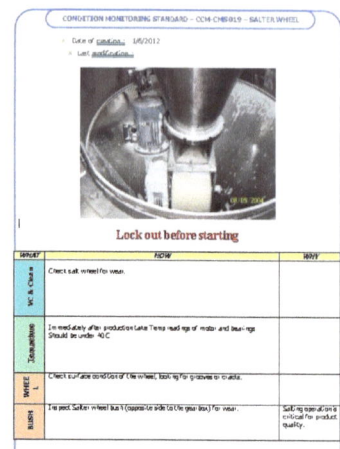

Figure 5.1 – Inspection Instructions

Infra-Red the New Frontier

Infra-red camera apps allow for quick and accurate inspections. Edward Demming advocated

"Compare one that works with one that doesn't." (Circa 1949)

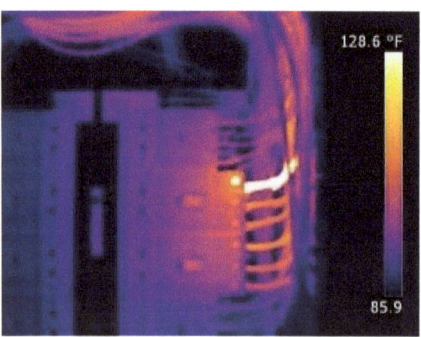

Mobile technology makes comparisons very easy. In this photo it is obvious that something is different.

The significance of the difference can be documented.

6.0 Cleaning

While it might seem a simple task, cleaning is critical to maintaining the effectiveness of many processes and equipment. The following items must be identified in an SOP's section on cleaning to ensure that a process is fully capable the next time it is needed.

6.1 Hazards When Cleaning
- List cleaning hazards including chemical risks
- List mechanical hazards
- List countermeasures
- Question whether the hazards can be eliminated by redesign

6.2 How to Clean
- Specific instruction on how things should be cleaned
- This can be documented by video and provided as training
- Isolation procedures that must take place prior to cleaning

6.3 What to Clean
- A determination on what should be cleaned
- Can be documented by video

6.4 When to Clean
- The frequency of cleaning is defined
- The reading and filling out documentation of cleaning will be covered in 2.0 of the SOP

6.5 Cleaning Schedules
- A Grid schedule is useful especially if MS Excel is used.
- Hyperlinks and comments can be used to train staff in the activities they are expected to perform.

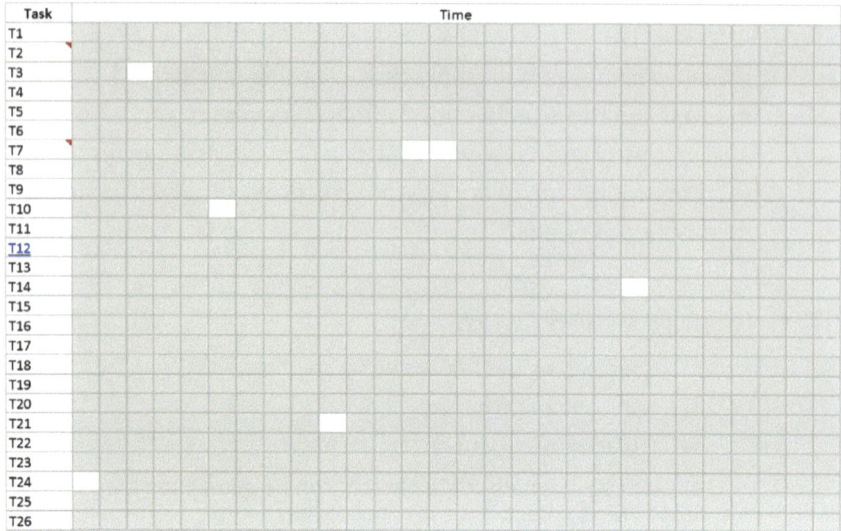

Here T12 has a hyperlink that refers to the appropriate quality procedure. There is also a comment on T7 to provide directions specific to that task.

6.6 Ask the Question
- Can this be done safer?
- Can this be done easier/better?

7.0 Inspection While Cleaning

What gets inspected while cleaning is performed? These critical items can be included on the cleaning schedule.

The importance of cleaning can be overlooked. Often it is done at the end of a production run and sufficient time may not be available.

Staff may not be trained how to clean and often resources are not provided to do so. Such an activity is simply expected, and leadership follow-up may not be a priority.

Examples of Good Cleaning Business Decisions:
$6,000 was spent on a floor cleaning machine.
Machine replaced hand mopping
Allowed the plant to be cleaned during production.
The machine paid for itself in labour costs in the first two months

Dry Ice cleaning was introduced
Cost was $20,000
Investment recovered in 2 months as electrical components were also cleaned.
Components were also cleaned.

What problems are we looking for?

- Unusual Smell
- Air/Hydraulic or Tank Leaks
- Walkway and Flow Blockages
- Temperature Change
- Broken Parts
- Worn Parts
- Missing Parts
- Missing or Loose Nuts
- Abnormal Movement in Equipment
- Vibration or Shaking
- Abnormal Heat
- Unusual Sounds
- Discoloration
- Use your Senses (Sight, hearing, smell, touch. You are a highly refined detection device)

Tips for Inspection:

- Don't stop with what we can see. Look for loose hardware, slight vibrations, higher than normal temperatures (don't get burned), and other problems that can be detected by touch.
- Make sure to check for worn pulleys and belts, clogged filters, grime on sliding surfaces, and any other problems that can cause premature failures.
- Do all gauges and sight glass work correctly, and are their functioning ranges clearly marked?
- Find sources of air and oil leaks.
- Are grease and oil locations conveniently located?

8.0 Preventative Maintenance

Preventative maintenance (PM) are activities that are completed to prevent or minimize breakdown. Often these are completed at start up or when cleaning.

Activities can include:

- Changing Filters
- Tightening Nuts
- Changing Oil
- Cleaning Filters
- Checking Belts
- Cleaning Belts
- Removing Dirt

Don't forget the office or lab:

- Printer Head Cleaning
- Instruments
- Computers
- Chairs
- Desks
- Doors
- Windows
- Air Con
- Writing instruments
- Pens

9.0 Predictive Maintenance

Predictive maintenance is methodology of predicting that the equipment is about to fail, by understanding mean time between failure (MTBF) or mean time between repair (MTBR). Its objective is to have equipment repaired before failure occurs

Some indicators are:
- Loss of Performance
- Increase Noise Level (Bearings)
- Increase Heat
- Vibration
- Increased Power Consumption (Motor Drives)
- Production of Infrared
- Ultrasound Production

Smartphone apps can be used to measure many of these parameters and make calculations that can help an operation avoid downtime created by unplanned equipment breakage.

To minimize the need for instruction, place a label on the plant with the range the operation should be in. This is the visual factory. All of these can be documented, and operators trained in their use.

Examples

Figure 9.1 – Predictive Maintenance Tools

Range Gauges

<u>Figure 9.2 – Range Gauge</u>

Gauges that indicate allowable operation limits can give a quick visual indication of whether a piece of equipment is ready to start-up or should be shutdown. Such indicators can show that a filter is clogged, which could be detrimental to a supply pump.

<u>Figure 9.3 – Visual Controls</u>

10.0 Theory

The theory of operation is to understand how a given process works. Troubleshooting, problem solving, and preventative maintenance instruction elements are needed to keep the process capable and in control.

10.1 Manufacturers Manuals

The original equipment manufacturer manuals are a great place to get the mentioned process information. Various applicable sections from the manuals can be pasted into the SOP to avoid loss of crucial information. The manual can be used as a reference document (i.e. Start machine by following section 97 of the manufacturer's manual).

10.2 Hand Drawn Process Map

Including a hand drawn or digital (MS Visio) process map can help the operation or maintenance personnel understand upstream and downstream equipment expectations. The map can also indicate how the rest of the system might suffer because of upstream or downstream. One of the best process maps to do illustrate system impacts is a value stream map. Companies like Opexperts.org can help you create and document your value stream map so that it can be included in your electronic SOP.

10.3 Effects of Material Defects

Document the effects on production of material defects. A great tool to do this is a process failure modes and effects analysis (PFMEA). The process map can be a great springboard for developing a successful PFMEA. The PFMEA can be linked to the SOP and read prior to production so all countermeasures are understood.

10.4 Specification Sheets

Many times, processes are required to produce a product or service within given specification limits. These specifications are often found on media such as part blue prints, medical regulation bulletins or company policy. Your SOP should reference any such documents. It is a best practice to ensure that as these specifications are updated, that your company has a procedure for making sure that the latest revision is what is referenced. It would be an awful situation to produce a day's worth of product with a capable process that has removed too much material, causing an enormous scrap cost.

11.0 Error Messages

At the back of manufacturer's manuals, is a list of on-screen messages.

- The small stoppages are usually more costly than the big dramatic ones
- Small stoppages are often very easily eliminated
- Another method is to get operators to compile their own list
- Every time a message appears, add it to a chart
- Excel is very good for compiling these and the messages can be sorted

11.1 Eliminating Error Messages

There needs to be a balance between eliminating error messages and fixing them. Measuring stoppages and restarts means that the cost of stoppages can be found.

- The small stoppages are usually more costly than the big dramatic ones
- Small stoppages are often very easily eliminated

11.2 Using Talley Charts to Measure Small Stoppages

Error message	Frequency
Too many starts	5 Occurrence
Low temp	3 Occurrence
Clean sensor	52 Occurrences

In this case the sensors were getting dirty most of the time. Installing a different sensor eliminated the problem

- Each stoppage took approx. 45 secs to fix
- There were 60 stoppages in two weeks
- Over 40 weeks it would be 1,200
- Downtime of 15 hours per year

If there are 10 similar stoppages, then total downtime for year could be 150 hr per year

12.0 Troubleshooting

"Alvin is our best engineer. When we have a problem, we call him and he will come and fix it. He knows exactly where to go and what to do."

His plant seldom ran for more than three hours without an issue

"Our Engineers are Lazy. They do not do anything. They just walk around and take samples and test the machines. Then, they get a contractor in to do the work. They do not do anything."

This plant had 99.95% reliability. On average there was one unplanned stoppage every 9 months.

Trouble shooting charts (Figure 12.1) are a two-edged sword. Often the time taken to document the fault and make such charts is more than the time it would take to eliminate faults.

However, for some processes a tool such as statistical process control can provide an early warning to allowing an operations to intervene early and avoid losing finished work. The example below shows data point 13 falling out of the process's upper control limit. This would be a good time to shut down and resolve the issue.

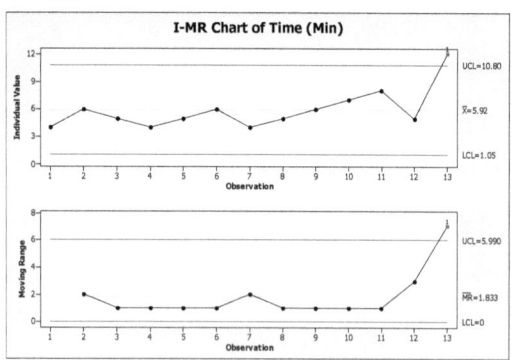

Figure 12.1 – Control Chart

12.1 End State Analysis

This is a most useful tool for trouble shooting. The end state analysis describes the faults that may occur in details and also lists the known solutions in order of likelihood.

Advantages of End State Analysis

Table is very easy to make

It is very easy to train to

12.1.2 Example of an End State Analysis Table

Machine will Not Start

1) Check That Power is "On" i. Plug another appliance into the same socket. Does it start? ii. If not, socket is faulty, call electrician	
2) Check On/Off Switch on Vac is On i. Check that switch is in the "on" position	
3) Is the problem intermittent? i. ⚠ If it is, disconnect machine and call electrician. ii. There is a risk of electrocution.	

Table 12.1 – End State Analysis Table

12.2 Manufacturers Manual.

Most manuals have troubleshooting sections on what to do to when there is a stoppage. Using this section is a good starting point. Many of these manuals are provided online and can be copied into the SOP or referenced using a hyperlink.

12.3 Influences of Specification Variation

One of the key elements that should drive process design and development of process SOPs are the specifications. Regardless of their type, specification generally have an allowable tolerance build into them. For instance, if you are producing a fastening nut, the thread diameter dimensions will be called out as 1.0 inches +/- 0.01". This allows the manufacturer to produce a quality part with a thread diameter of 1.01" or 0.99".

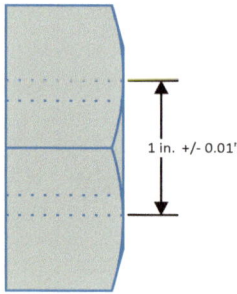

Figure 1.3 – Allowable Variation Example

The allowable variation must be well understood when developing a standard operating procedure. Issues that arise with providing a product or service must be visible to the operator of the process. Therefore, the SOP should not only reference these tolerances, but should build predictive maintenance and safety stances around their ability to produce at the specified tolerance.

12.4 Six Sigma Problem Solving

12.4.1 Belt Levels

White Belt Level
Small practical improvements. Typically, these are improvements suggested by staff. The thrust is to remove variations in the process.

Green Belt Level
SOP and Lean tools introduced. A more structured approach to problems is easier because there is less variation in the system.

Black Belt Level

Black Belt Level
A highly structured approach to problem solving is underway. Team meetings and high levels of communication between all parties is introduced.

13.0 Training & Assessment

13.1 Training Screens

- If there are training screens, these should be documented in the SOP and linked as a controlled document.
- How to make simulated training screens are covered in another of my publications

13.2 Making Training Structure

The SOP can be broken up into trainable sections that can be systematically introduced

Stage One

1.0 Safety
2.0 Record Keeping

These can be trained before Trainees start on the plant. It also removes "Background noise." Specific training safety training is introduced at the time trainees learn the parts of the process.

Stage Two

3.0 Starting
4.0 Operating
5.0 Shut down
6.0 Cleaning

This covers the basic operation of the plant.
If the Stage one is trained out, then it is easier for trainees to learn- fewer distractions

Stage Three

7.0 Inspection while cleaning
8.0 Preventative maintenance
9.0 Predictive Maintenance
10.0 Theory
11.0 Error messages
12.0 Trouble shooting
13.0 Training plans and Training aids
14.0 Revision Log
15.0 Appendix

Advantages of Grouping These Headings at the End:

Removes distraction
A goal is to develop understanding, from which trouble shooting can be developed
It is always better to place the theory on top of the practical, so theory has a bed to sit on.

13.2.1 Example of a Training Sheet

Kalamazoo Stage One: Making a training Plan NZQA 1289

Name_____ Date / /

SOP Section	Page	Trainee	Trainer
1.0 What is an SOP			
1.1 Links to Inductions			
1.2 The SOP Structure			
1.3 Embedding Safety			
1.4 Embedding Quality			
1.5 The Place of 5S			
1.6 Training Within Industry.			
1.7 The Questions that Must be Asked			
2.0 The mechanics			
2.1 Any Topic			
2.2 Snipping Tool		`	
2.3 Windows Steps Recorder			
2.4 Print Screen (Copy Screen)			
2.5 Video Screen Capture			

Table 13.1 – Training Sheet Example

13.3 Assessment

The training sheet documents that trainees have been trained in the procedures. Next, they will need to be assessed to determine if they have the knowledge and understanding. This assessment is critical if we are using video for training. The assessment should be given at least two weeks after training has been delivered to ensure information retention.

It will also determine if trainees are using the SOP for reference purposes.

An acceptable response from trainees to be in the practice of saying, "I will check it in the SOP".

There is a difference between remembering something and being able to locate knowledge.

13.3.1 What is Important

A) Select What is Important to Know i. Select what is essential for Trainee to know	
B) Determine What is Important to Find i. Determine what trainees need to be able to find. ii. This are things that need not be remembered provided trainees can demonstrate how to obtain information	
C) Understanding i. What understandings do the trainees need to demonstrate e	
D) Demonstrate Proficiency i. What do trainees need to demonstrate they can do	
5) Link to Unit Standard i. The assessment should be to a department standard. ii. Performing the assessment will then meet departmental standard.	

Table 13.2 – Training Sheet Contents

13.3.2 Example of Assessment Sheet

Kalamazoo Stage One: Assessment NZQA 1289

Name _____

Date _____

Assessor _____

Question	Answers: 1) These are taken from the SOP 2) Print answers here
1) Known Hazards What are the known Hazards in the area?	Refer to XXXXX

What PPE Needs to be Worn
  1) Is PPE always Effective?
  2) Why?
  3) What does PPE guard against?
	Refer to XXXXX

3) What Records Need to be Kept i. Provide copies of correctly filled out records	List the records and provide copies

Table 13.3 – Assessment Sheet

13.4 Linking to Departmental Standards

This is an extract from a NZQA Standard. Green writing covers how the evidence requirements are covered by the SOP

NZQA 1289 ---- Operate a chainsaw and carry out basic chainsaw maintenance in a commercial forestry situation
Level 3   Credits 12

Evidence Requirements

A.   Safety features are identified, and their main function is explained in accordance with the Best Practice Guide. Range on/off switch, throttle lockout, chain brake, rear hand-guard, spark arrester/muffler, anti-vibration mounts, chain catcher, mitt. Covered by H/S section and Assessment Q1

B.   Chainsaw starter cover/assembly, top cover, air filter, side cover, bar and chain are removed in accordance with chainsaw manufacturer's recommendations. Covered by 3.1 which is a copy of the manufacturers manual, and assessment question Q11

C.   Chainsaw components are identified, and their main functions are explained. Range starter mechanism, flywheel, cooling vents, top cover, air filter, carburettor, spark plug, high-tension lead, cooling fins, choke, throttle, side sprocket, chain tension adjuster, clutch, bar oil hole, bar rails and groove, bar sprocket, depth gauge, drive link, cutter, rivet, tie strap. Covered by 3.2 which is taken from manufacturers cover, drive manual and assessment Q12.

14.0 Revision Log.

The revision log should contain:

1) SOP Owner
2) Current version number
3) A list of previous versions
4) A list of documents the SOP refers to

Section (Identify SOP Section & page)	Date of Change	Description of Change (describe what has changed)	Old versions removed?
1	12/19/29		
2			
3			
4			
5			
6			
7			
8			
9			

Table 14.1 – Revision Log

15.0 Appendix

Item	Where found
Carton supplier	
Filters	
Ink cartridges	
Maintenance	
Printer Rep	
Quality Sup	
User manual	

Table 15.1 – Appendix

This is where forms images where to find things Contact details are found. It can have its own table of contents

Opexperts Consulting Company LLC has been mentioned several times in this publication. The company's owner and co-author of this book, has worked with and consulted many organization who learned the importance of standard operating procedures. Many of these companies had an "ah-ha" moment after going through extensive process improvement changes, learning that they didn't have a mechanism in place to solidify the gains they had made.

Opexperts is happy to help you develop standard operating procedures as a one-time engagement. We can also help you establish an organizational standard, which gives your staff the tools needed to make development and maintenance of SOPs a standard practice of the business.

Contact us for this and any other process improvement need you might have. We've even recently gained capability to help you put establish a thought leadership presence with professionally filmed content for your website. Our specialty is helping business improve at a comfortable pace and cost.

Contact Opexperts Today:	
Phone:	(214)334-8293
Email:	info@opexperts.org
Web:	opexperts.org

www.ingramcontent.com/pod-product-compliance
Lightning Source LLC
Chambersburg PA
CBHW040228220526
45473CB00001B/162